Printed in Mexico

ISBN 978-0-15-362265-6
ISBN 0-15-362265-2

2 3 4 5 6 7 8 9 10 050 16 15 14 13 12 11 10 09 08

Visit *The Learning Site!*
www.harcourtschool.com

Natural Resources

We get many of the things we need to live from Earth. We get oil and gas to heat homes and drive cars. We get rocks and minerals to build buildings and roads. We get trees to build houses. We get water and plants that are the basis for all life and energy. Earth has an abundance of natural resources. A **natural resource** is any material that occurs in nature that is essential or useful to people. Some of these resources, such as water, are necessary for life. Others, such as diamonds, are simply useful and enjoyable.

Fast Fact

The oldest rocks on Earth's surface are almost 4 million years old. They are called Acasta gneiss and are found in Canada.

The ocean supplies us with many natural resources. People all around the world depend on fish for their food.

How does your community use natural resources? It probably uses some of the trillions of kilograms (billions of tons) of gravel, stone, and sand that are mined each year. These natural resources are used to pave roads and build sidewalks, skyscrapers, and other buildings.

All communities use oil and gasoline. Throughout the United States each person uses about 4,450 liters (1,180 gallons) of oil in just one year. Resources such as iron, tin, aluminum, and other metals are used to build everything from bridges to fences and to make everything from doorknobs to soda cans. Even in school, the paper products, wooden pencils, and books you use are made from trees, another natural resource.

Focus Skill **MAIN IDEA AND DETAILS Select one natural resource. Give examples of how this resource is used in your community.**

Most of the world's gold, a valuable natural resource, is found in a country called the Republic of South Africa.

Digging Deep

Oil is one of the most valuable natural resources. Crude oil is found between layers of rock deep beneath Earth's surface. It took millions and millions of years to form this oil. Crude oil is made from decayed material. Tiny sea organisms that died millions of years ago were buried under layers of soil and rock. Over the years the rock and dirt piled higher and higher. This caused increased temperature and pressure on the organisms. Over time, this caused the organisms to turn into crude oil. Because the oil formed from the buried remains of once-living organisms, it is called a **fossil fuel**.

People and machines drill deep into Earth to get crude oil. Wells and pumps bring the oil to the surface. Once it reaches Earth's surface, it's taken to a refinery. The refinery "cooks" the crude oil into the products people use every day.

Oil is valuable because it is the source of so many things people use every day. This oil well is being used to bring oil up to Earth's surface.

Peat is "young" coal. Heat and pressure are just beginning to form fossil fuel from decaying organisms. Even at this early stage of fossil fuel formation, peat can be used as fuel.

Coal is also a fossil fuel. Coal was formed from trees and other plants that grew in huge swamps a long time ago. Some coal lies close to Earth's surface and some coal lies deep under the surface. Today, coal miners use huge machines to dig deep tunnels. Some miners will dig as far down as 610 m (2,000 ft) into Earth, while other miners dig on or very close to the surface. Coal that has been forming for millions of years is harder than the "younger" coal. Peat, found on Earth's surface, is the very beginning of the formation of coal.

Fast Fact

Some buildings are heated with geothermal energy. This heat energy comes from hot, molten rock deep inside Earth that flows near Earth's surface.

MAIN IDEA AND DETAILS **Why do we look under Earth's surface to find important fuels?**

More Gifts from Earth

Earth's forests also provide natural resources. Trees are used for paper products as well as lumber. Instead of taking trees from existing forests, many companies now plant tree farms for these products. When large trees are cut down, new ones are planted in their place. This way there will be new trees to replace those that have been cut down for lumber or other uses.

Trees are planted to replace the trees that are used. This tree may take thirty to seventy years to grow.

The Hoover Dam in Nevada turns the energy from moving water into electric power.

Where would every living thing on Earth be without water? Water is another natural resource. People in communities around the world need water to drink, cook, bathe, feed their animals, and water their crops. Water can also be used to run electrical generating plants.

Soil is also one of Earth's natural resources. Plants that grow in the soil use the sun's energy to make their own food. Those plants then become a food source for all living things. People around the world make use of the animals that eat those plants. Animals such as cattle provide food and milk, as well as leather for clothing.

MAIN IDEA AND DETAILS Explain ways in which soil and water are important natural resources.

Will They Run Out?

How do you think people would be affected if these natural resources ran out? How can people learn to use them in the best way? Fossil fuels took millions and millions of years to develop. Once people use up the oil and coal, it will be gone. That's why these fossil fuels are called nonrenewable resources.

Trees can be planted over and over again. Seedling trees can grow and form new forests. That's why trees and lumber are renewable resources. However, some animals and plants thrive only in forests that have very old trees. Once the very old forests are cut down, they are gone for a very long time because trees don't grow very fast. Even though replanting new saplings will renew the supply of lumber, the living things that depend on old growth forests will suffer if those forests are cut down.

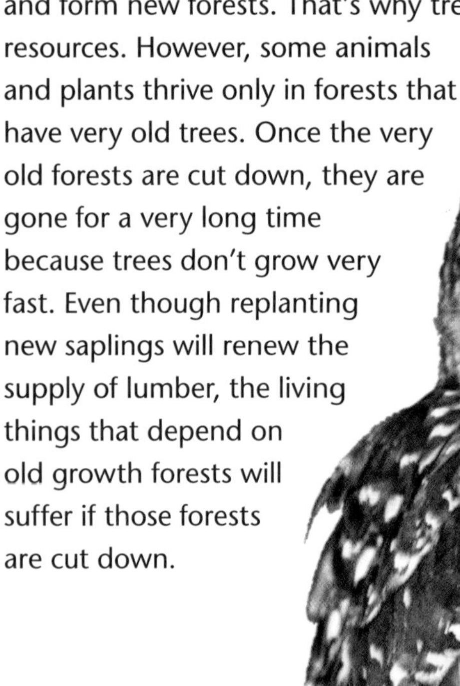

The spotted owl is rapidly decreasing in population. Spotted owls need large areas of old growth forests to survive and many of these are being cut down.

Renewable resources can be replaced within a human life span. Corn, wheat, and cattle are considered to be renewable resources. Farmers grow and replace these food sources over and over again. Wood that is burned for fuel is gone forever once it's used up. But because wood can be replaced with new seedlings that will grow into trees, wood is a renewable resource.

Fast Fact

Even though the plants and animals from which fossil fuels were formed were buried millions of years ago, they still contain the energy they received from the sun.

Some things are called reusable resources. They can be used over and over again. For example, there is only a certain amount of ore in the ground that can be made into aluminum. That ore is a non-renewable resource. But by recycling and reusing aluminum cans, people can make sure that aluminum is a reusable resource.

SEQUENCE **Describe the sequence of events for a renewable resource such as corn, wheat, or lumber.**

Conservation

Imagine being on a camping trip. Your knapsack falls into the water and travels downstream. All your food is gone. The only thing you have to eat is one apple and it has to last you for the whole weekend. How will you make it last? You will probably eat just a little bit at a time, saving some for later.

This is conservation—using resources carefully so that they'll last as long as possible. People have to practice conservation with many natural resources. The nonrenewable resources can't be replaced within a human lifetime, so it is wise to practice conservation of these resources.

Fast Fact

Cameras, backpacks, and even some clothing are products that might be made from recycled plastic.

When forests are cleared, it can take decades for trees to grow back. Practicing conservation saves trees.

This cassowary is a non-flying bird that stands nearly 5 feet tall! It lives in bauxite-rich areas of Australia and depends on fruit that falls from trees. These trees can be destroyed when land is mined for bauxite, which is used to make aluminum. Reducing, reusing, and recycling help protect this endangered species.

One way to do this is to reduce the amount of resources used. A simple thing such as riding a bike or walking instead of driving a car can reduce the use of oil and gasoline.

Another way to conserve is to reuse something. Many resources are used up to replace the paper and plastic bags that get thrown away when you return from a store. But if you take a canvas bag to the store you can reuse it many times. This will save natural resources.

Recycling cans, bottles, plastic, and newspapers is another good way to practice conservation.

CAUSE AND EFFECT **Explain how recycling can help animals in the rain forest.**

People and Ecosystems

People need resources to live. But people need to be wise in gathering and using these resources.

Burning coal releases sulfur dioxide, which is a contributor to the production of acid rain. All around the world, acid rain has destroyed plant life and has killed off fishes in lakes and rivers. Today, many governments have programs to reduce acid rain by cleaning factories and limiting the output of sulfur dioxide.

Burning fossil fuels such as oil and gasoline can cause air pollution, which is harmful to both plant and animal life. To help with this problem, cars and other machines dependent on fossil fuels have been developed to produce less pollution. Also, alternative energy sources—from the sun, wind, and water—help make people less dependent on the use of these resources.

People have helped the environment by writing laws that limit the amount of pollution factories and cars can release into the air.

The act of simply transporting oil can be dangerous. Boats that carry oil can leak. In the waters off Alaska in 1989, eleven million gallons of oil spilled from a single tanker, the Exxon Valdez. Oil-polluted water killed fish. Lumps of oil washed up on shore, killing other animal life. Since then, regulations have changed. Oil tankers must have double hulls. Escort ships and specially trained marine pilots watch over the tankers. Special equipment has been developed to help clean waters quickly in case a spill still happens.

Fast Fact

In the 1300s people in Europe feared and killed cats. As a result, the rat population increased. The rats carried fleas that caused a plague. Millions of people died.

Wolves have been hunted and have nearly disappeared in some parts of the United States. Scientists have worked to reintroduce these predators.

People can also affect ecosystems by introducing animals and plants to an environment in which they aren't naturally found. To attract tourists, Nile perch were once introduced to several large lakes in Africa. But the perch were predators. They killed off many of the fish that were native to these lakes. Today, wildlife managers and government officials are careful to protect the natural balance in ecosystems.

CAUSE AND EFFECT What are ways in which the actions of humans can help protect the environment?

Endangered Lives

Homes, shopping malls, and roads are built every day to meet the needs of the world's growing population. Human development constantly moves into areas that were once tropical rain forests, prairie land, grasslands, marshes, and estuaries. An **estuary** is an ecosystem where salt water and fresh water environments come together. Many fish hatch in estuaries. Estuaries can suffer from pollution, land development, and too much fishing.

In the process of human development, animals and plants can lose their habitats and water sources. Animals and plants that need these specific ecosystems can become endangered or even extinct. An **endangered species** is one that is close to dying out completely. **Extinction** is the loss of a species forever.

Fast Fact

The Florida Everglades is the only place in the world in which crocodiles live alongside alligators.

Government agencies work to protect these important ecosystems.

Wild animals such as mountain lions, moose, and deer often have to look for food in populated areas. Their natural forest habitat has been replaced by people and houses.

Wetlands are an example of another fragile ecosystem. A **wetland** is an area of land that is usually covered with water. Wetlands are natural filters that clean pollutants out of the water. They're home to many animals such as frogs, alligators, and wading birds.

In the past, many of them were drained for construction. Many of the animals living in wetlands became endangered species. People didn't realize the value of wetlands. Now these valuable ecosystems are watched over more closely.

CAUSE AND EFFECT **What might be the result of an estuary becoming polluted?**

The Everglades in Florida is an important wetland. Although much of it was once drained to build farms and homes, restoration efforts are now underway.

Summary

All living things share Earth's natural resources. Everything is connected, and changes can help or harm many things throughout an ecosystem. Now you know that you, too, can help in your own way by being aware of the problems and practicing conservation.

Glossary

endangered species (en•DAYN•jerd SPEE•sheez) A species with so few individuals that it could die out (14, 15)

estuary (ES•choo•ehr•ee) The place where a freshwater river empties into an ocean (14)

extinction (ek•STINGK•shuhn) The loss of an entire species (14)

fossil fuels (FAHS•uhl FYOO•uhlz) An energy-rich resource formed from the buried remains of once-living organisms (4, 5, 8, 12, 13)

natural resource (NACH•er•uhl REE•sawrs) A material that occurs in nature that is essential or useful to people (2, 3, 4, 5, 6, 7, 8, 10, 15)

wetland (WET•land) An area of land that is covered by water all or much of the time (15)